WIND
POWER

Designed and produced by
Aladdin Books Ltd
70 Old Compton St
London W1

First published in the United States 1985 by
Gloucester Press
387 Park Avenue South
New York, 10016

Library of Congress
Catalog Card No. 85-70600

ISBN 0531 170071

Printed in Belgium

The cover photograph shows a typical winter landscape in the Netherlands.

Photographic credits:
Cover and pages 6 and 9, Zefa; page 4/5, Daily Telegraph; pages 7 and 18/19, Earthscan; pages 11, 13 and 15, Friends of the Earth; page 17, Sandia Laboratories; page 21, Coventry (Lanchester) Polytechnic; page 22/23, Gifford Technology; page 25, Bruce Colman.

ENERGY TODAY

WIND POWER

MIKE CROSS

Illustrated by
Ron Hayward Associates

Consultant
Stewart Boyle

Gloucester Press
New York : Toronto

Introduction

Our lives depend on energy. We need it to heat our homes, to make electricity and to move us around on land and in the air.

The wind is a turbulent, erratic force but we can actually harness it as a source of energy. For centuries ships and windmills have successfully used the power of the wind.

Today we are looking for new ways of using the energy of the wind and even the waves. The technology is often costly but the advantage is that they are both free, and will not run out.

Natural power in a high wind in Brazil

Contents

Energy from the wind

Windmills – like the one below in Crete – have provided power for hundreds of years. But while they can pump water and grind corn, they are too small to make large amounts of electricity.

Engineers are now testing new "wind turbines," like the ones opposite. These are much more powerful than windmills. The hope is that they will soon provide a major extra source of electricity, and that people all over the world will benefit from them.

Cretan windmill

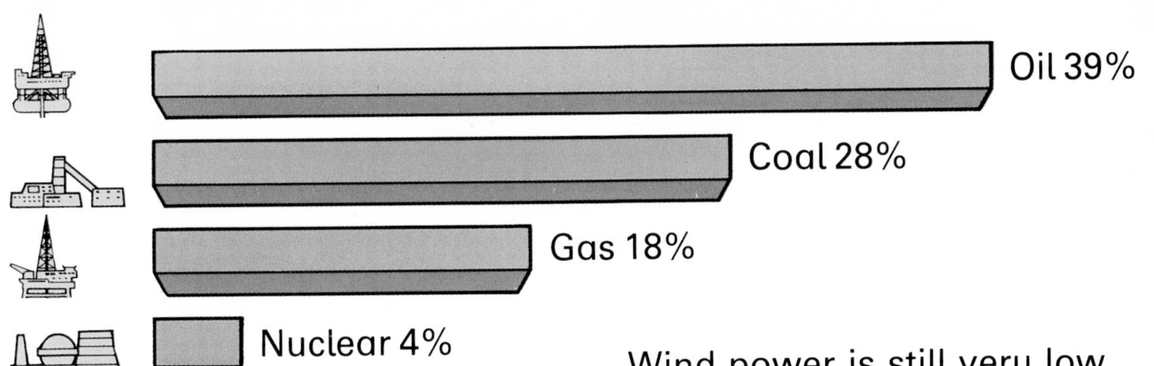

Wind power is still very low on our table of energy consumption. But as the major fuels gradually run out, wind power must become more widely used. Its greatest importance will be in poorer countries, where even firewood is becoming scarce.

American wind turbines

Catching the wind

Traditional windmills and modern turbines work roughly the same way. Both need a device to make sure that the blades always face the wind. In a windmill there is usually a "vane" or fantail. In the largest turbine there may be a computer-controlled motor.

When the wind blows the vane rushes away to avoid it. This action makes the part of the mill with the blades swing around into the wind and they start turning. They turn because they are shaped to work *with* the wind rather than against it. They are curved – like the wings of an aircraft – and sometimes they are tilted at an angle against each other.

The wind blows, hitting the fantail (1). The fantail rushes "around the corner" to escape the wind, shifting the position of the mill. The louvered blades, (2), accept the wind (3), and turn, driving the spindle and machinery inside (4). If the wind shifts, the fantail will move the mill until once again the wind and blades are working together.

The blades of both windmills and turbines must be able to adjust to the wind speed. If they cannot do this, strong gusts may damage them. The old mill in the photograph would have had "louvers" on its blades, rather like louver blinds. These were opened and shut according to the wind speed.

Nineteenth-century windmill, England

Savonius rotor

The "Savonius rotor" shown in the photograph is a wind machine made from very cheap materials – an old oil drum cut in half and mounted on a wooden frame. Farmers use it to pump up water and irrigate fields.

It is cheap and easy to make, and is used in the developing countries in Africa and Asia. If the machine breaks down it can be repaired in a village workshop – expensive parts do not have to be imported from abroad. But while the Savonius rotor is useful for pumping water, it is not powerful enough to produce electricity.

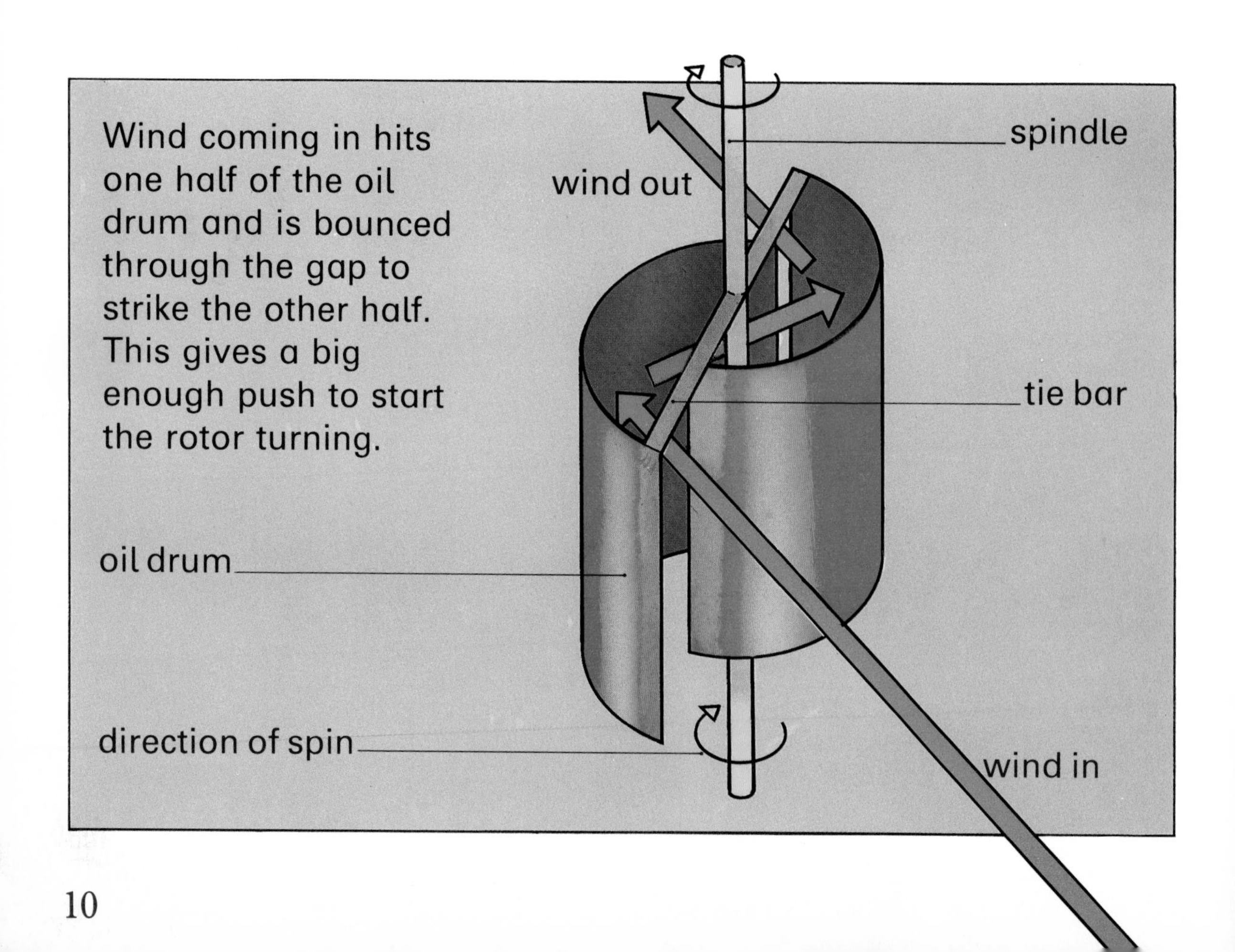

Wind pumps

Wind machines have been developed in different ways around the world. In North America engineers discovered that many small blades were better at catching the wind than four large ones. Wind pumps, like the one in the photograph, were installed all across the Midwest. Farmers could now pump water up from very deep wells, and the land was opened up to cattle ranching.

This type of wind pump can still be seen on many farms in the USA and Canada. There are more than one million around the world.

Some farmers, though, have given up their wind pumps and replaced them with pumps powered by diesel oil or electricity. The reason for this is that the wind does not always blow when it is needed!

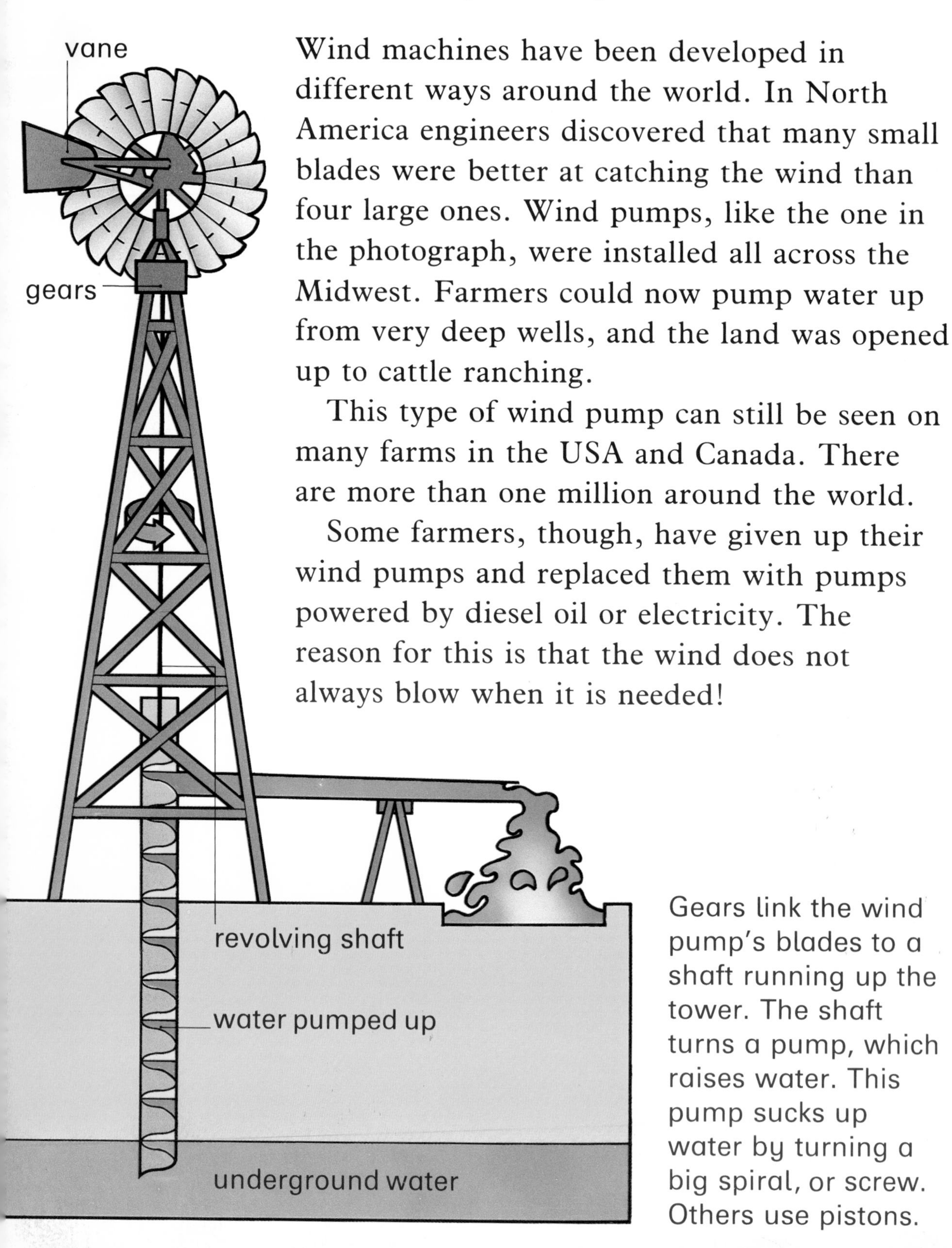

Gears link the wind pump's blades to a shaft running up the tower. The shaft turns a pump, which raises water. This pump sucks up water by turning a big spiral, or screw. Others use pistons.

Wind turbine in the Orkneys, Scotland

Wind turbines

Nowadays, though, one of the best ways of using the wind's energy is to turn it into electricity. Some small communities, and even some homes, now have their own small turbines to do this.

But some wind turbines are very large. The one in the photograph is 46 meters (150 feet) high. It has slender, shaped blades which are controlled by computer. The tips of the blades can swivel, and they act as brakes in strong winds. The top part of the turbine houses all the controls and swings to face the wind.

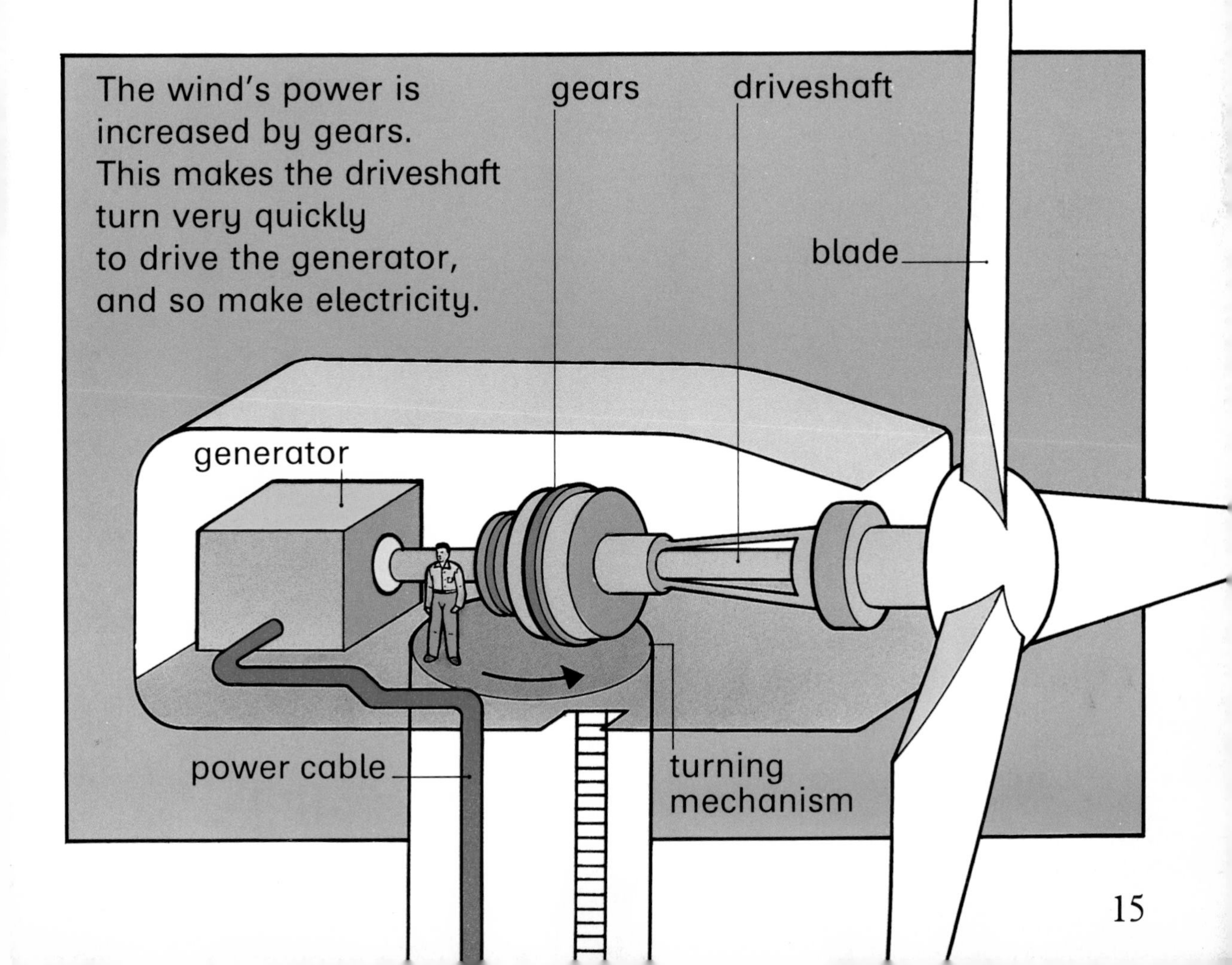

Darrieus

The strange machine pictured here is sometimes called an "eggbeater" because of its shape. The proper name is the Darrieus wind turbine, after its inventor.

These turbines have a vertical axis, like the Savonius rotor, but they are much more powerful. Several kinds of Darrieus turbine are being tested in the USA and Canada; a very large turbine, which could generate up to one megawatt of electricity, is planned. This turbine could cut the present cost of producing electricity from the wind by almost half.

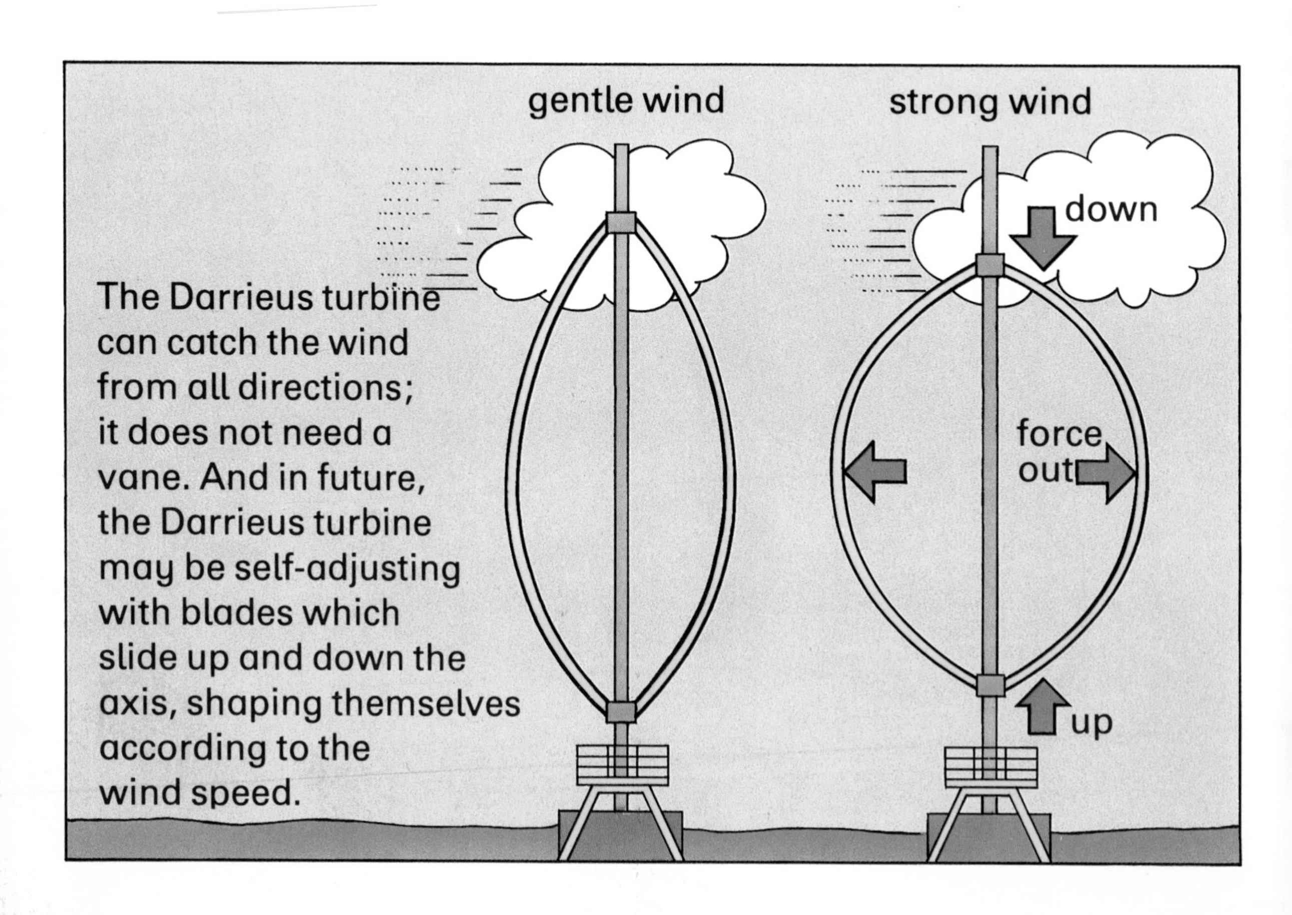

The Darrieus turbine can catch the wind from all directions; it does not need a vane. And in future, the Darrieus turbine may be self-adjusting with blades which slide up and down the axis, shaping themselves according to the wind speed.

Wind farms

Sometimes groups of wind turbines may be set up in one place. This is called a "wind farm."

Wind farms have existed around the Mediterranean for centuries; now modern ones have been started in the USA. Wind farms sound like a good idea, but the power they produce is still erratic because the wind blows in gusts. A few people think they are ugly and noisy, but the land where the wind farms are located is quite often remote, of poor quality, and of little other use.

Wind farm at Altamont Pass, California

On windy days, turbines generate much more electricity than on calm days, and this extra electricity may not always be wanted on the grid. It can be "stored" by using it to heat up reservoirs of oil or water, but this is wasteful.

Batteries for storing massive amounts of power are now being developed. They have to be very big, or many small ones have to be used together. One large battery already being tested may be able to store 100 megawatts of heat.

Sea Clam

At sea, the wind makes waves – yet another source of free energy which we can harness. And waves are a more constant source of energy than the wind because they keep rolling long after the wind has died down.

At the moment getting energy from the waves is very expensive, because of the technology involved. But eventually the rising cost of oil should make it more realistic.

The turbine blades spin to produce electricity in the generator. Then the electricity passes into the cable, which links all sections of the Clam, to be collected on shore.

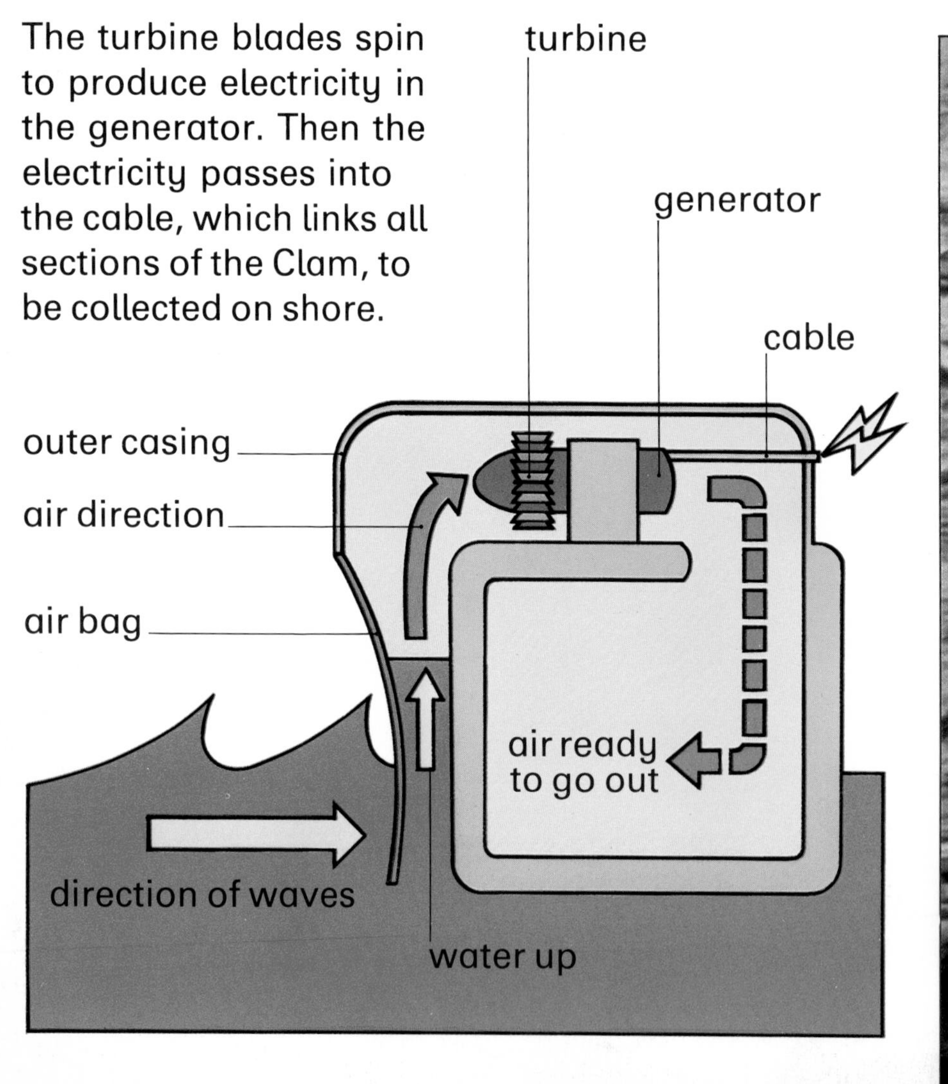

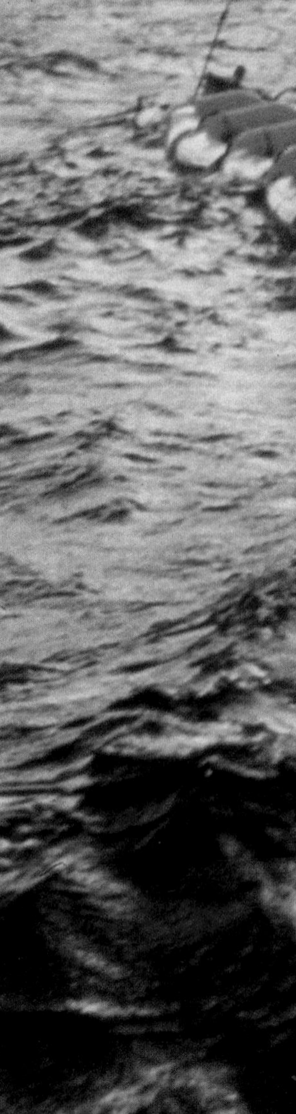

The Sea Clam, below, is a long barrier which can turn the action of the waves into power. Each of its sections contains a huge rubber air bag and a turbine. As a wave passes the bag is squeezed and air is pushed up and through the turbine, spinning the blades. When the Sea Clam is in the hollow of the wave, the bag is sucked out and air rushes back through the turbine to fill the bag again.

Sea Clam trials, Loch Lomond, Scotland

Wave rafts

Nobody knows how much energy is locked up in the oceans. But each wave that crashes onto the shore carries an energy potential of ten kilowatts in every square meter.

It takes clever engineering to turn the up and down motion of the waves into power that we can use. The "wave raft" on this page is still experimental, but it works. The raft is really two hinged platforms. As they rock up and down, the movement is converted into electricity by a generator on the large platform.

Wave raft in the Solent, UK

In the future, wave rafts might be moored off the coastline. The electricity produced would be gathered up in a central collection point and sent ashore by cable.

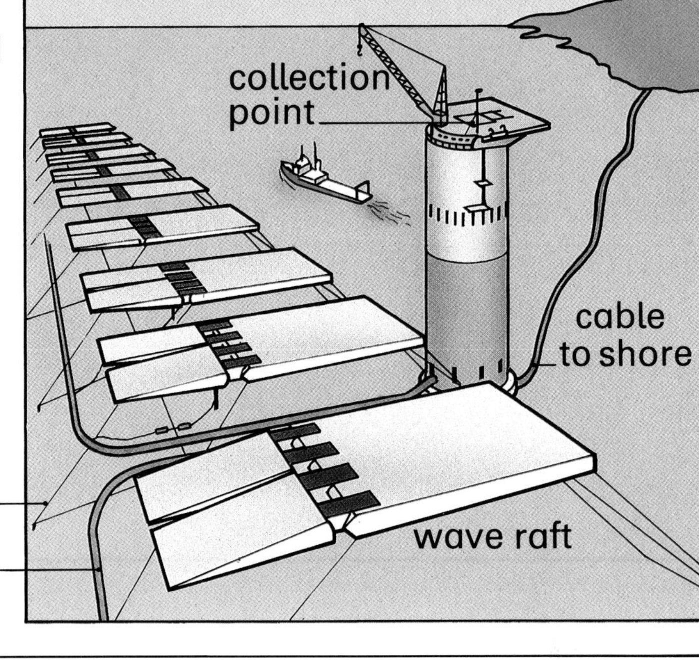

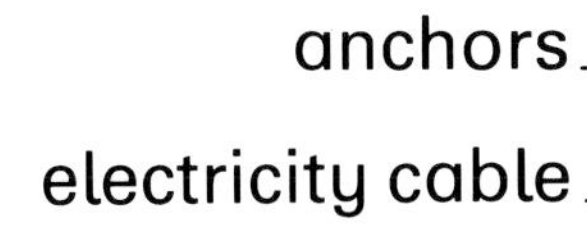

Sail power

For thousands of years sailing ships used the wind to move them along. But with the coming of steam, and then diesel oil, sailing ships became a thing of the past.

Shipbuilders are now rediscovering sail power and ships are once again being fitted with sails. This Japanese oil tanker has two rigid, plastic sails which can take advantage of the wind and so save precious fuel.

Sail power is ideal for ships which carry bulk cargoes. They use lots of fuel and they can make great savings by using the power of the wind.

The sails are linked to the ship's engines by a computer. They can be turned to catch the wind when it comes from the side, and folded up in strong winds.

"Shin Aitoku Maru"
in the Sea of Japan

Fact file 1

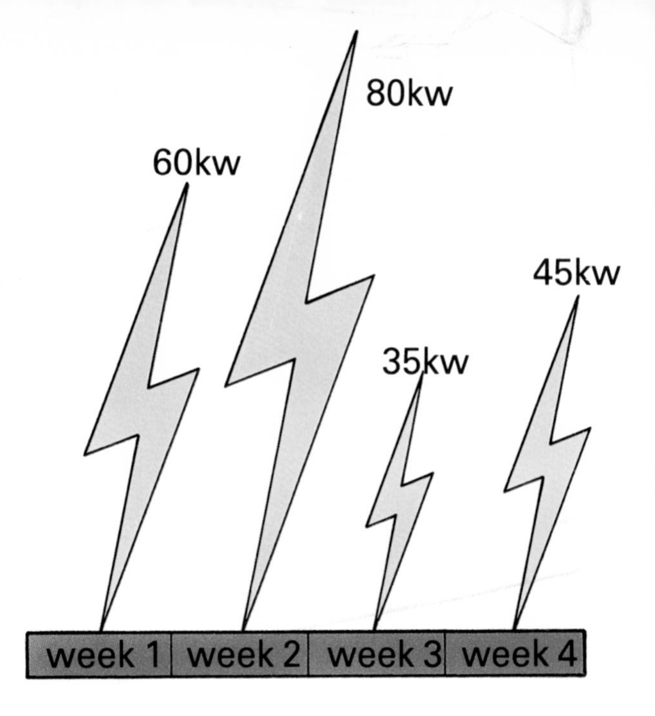

The diagram (right) shows that the wind is a variable source of power. It illustrates the output of a turbine with a generating capacity of 200kW over a month. Output varies – in the best week it was averaging only 80kW. Even so, certain winds across the world do blow at regular times in the year, and turbines do take advantage of them.

Wind power may be variable and surpluses difficult to store but it still makes sense. Because every time we use wind power, we save on precious fossil fuels.

The map below shows the world's windiest places, which include coastlines and islands. If, and when, wind power is developed on a large scale, we will know where to put the wind turbines!

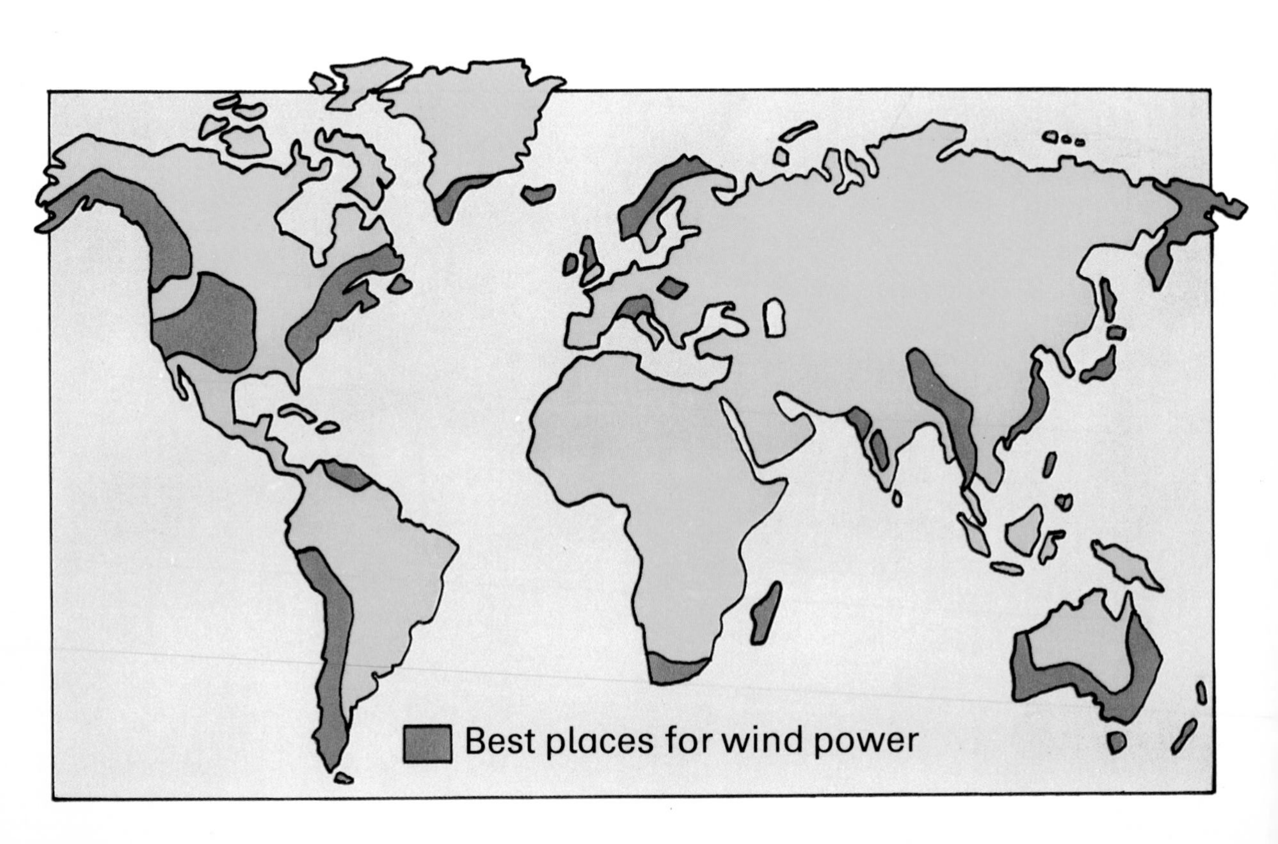

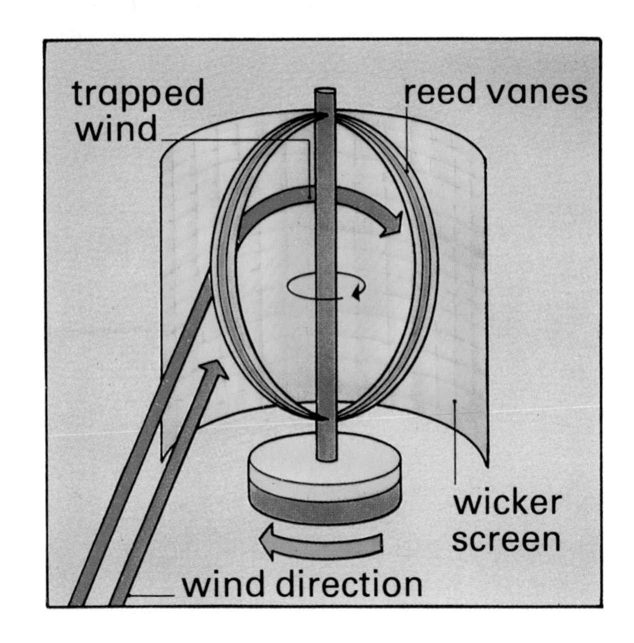

Windmills were first used in the Arab world and the Middle East. The one in the diagram dates from sixth-century Persia. It had screens to trap the wind and is a little bit like today's rotors.

The heyday of the windmill was the nineteenth century. Britain and the Netherlands each had 10,000 mills, and there were 1,800 in Germany. They were used for milling and grinding crops, and for pumping water.

A single windmill – with blades measuring 25 meters (82 feet) – could do the same amount of work as over 250 people. This would produce 240kw hours of energy in one day. A hard-working laborer can usually only manage about one kw hour of energy a day. We can see how important windmills were before steam, and later diesel oil, came into use.

Air conditioning was invented in the Middle East, where fans and screens have been used to cool houses for centuries. Some homes in Iran, Pakistan and Dubai have wind "scoops" mounted on the roof – like the one in the diagram. The scoops face the direction of the seasonal winds, drawing cool air into the rooms below.

Today, the wind's energy can even help plants grow! In some places, wind-powered heaters keep greenhouses warm.

Fact file 2

At the moment getting energy from the wind costs three times as much as getting energy from the fossil fuels: oil, coal and gas. But wind power has a great future if we can improve the technology – and it is being improved all the time.

Also, wind turbines are almost pollution free. They do not create smoke or acid rain. They are expensive to build, but easy to look after.

Three huge turbines, called Mod 2s, have been set up in the USA, in Washington State. 100 meters (330 feet) high, they can be seen from as far away as 8km (5 miles).

While tall turbines are effective at making power, they are subject to buffeting from the wind. The diagram below shows some different types of wind turbines and their size in comparison with a jetliner and an electricity pylon.

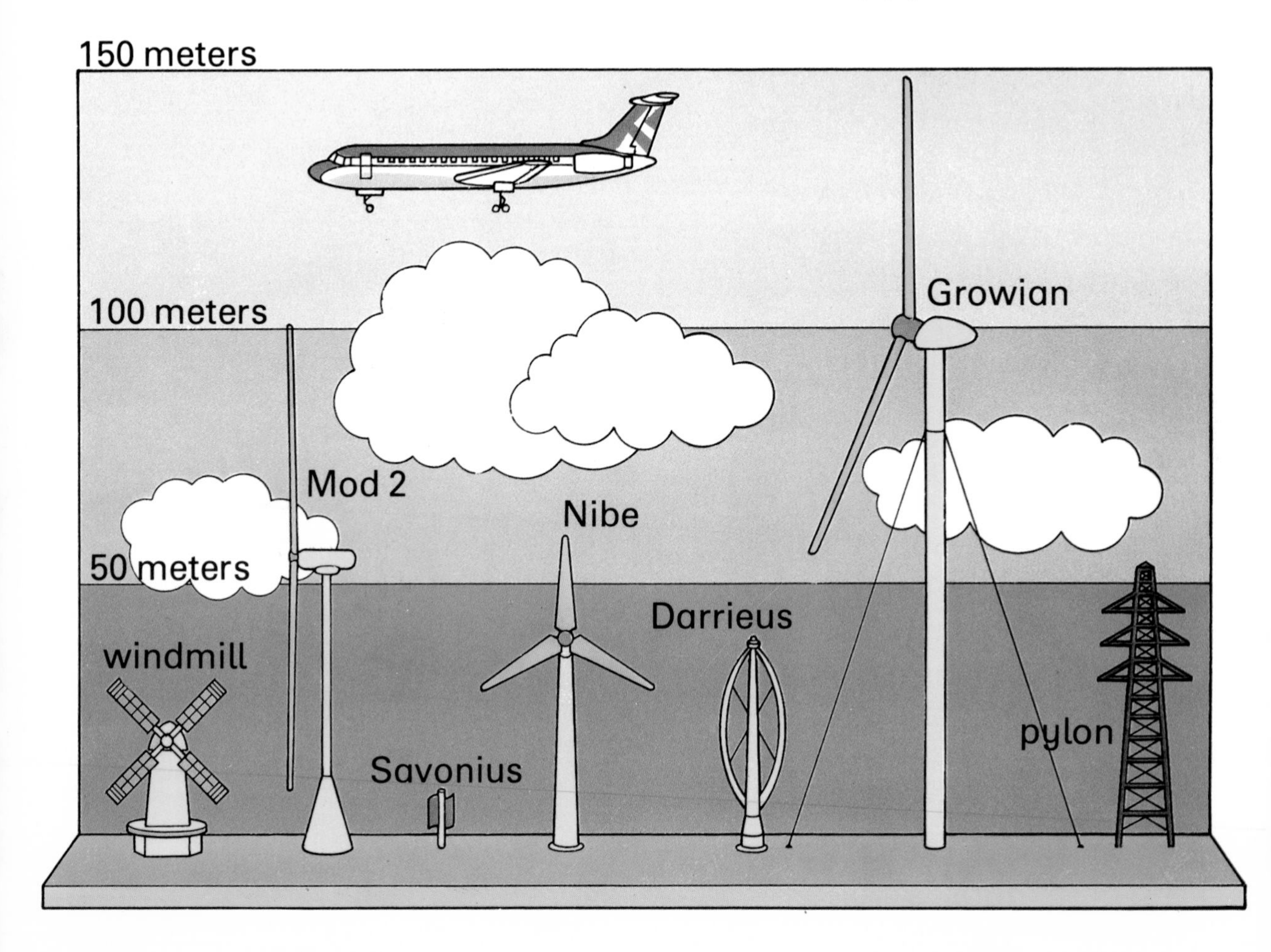

The USA has become one of the world's leading producers of wind energy. Some state governments give tax relief to people who set up wind turbines or invest in wind farms.

The UK has some big wind turbines sited in the Orkney Islands – one of the world's windiest places. British engineers are also eager to set up wind farms out at sea.

Sweden has several wind turbines. Two large turbines – similar to the Mod 2s – have been set up as an experiment; they are being studied to see which one works best.

West Germany has the largest wind turbine in Europe. The "Growian" has rotor blades 100 meters (330 feet) in diameter. The name Growian means "Big Wind plant" in English.

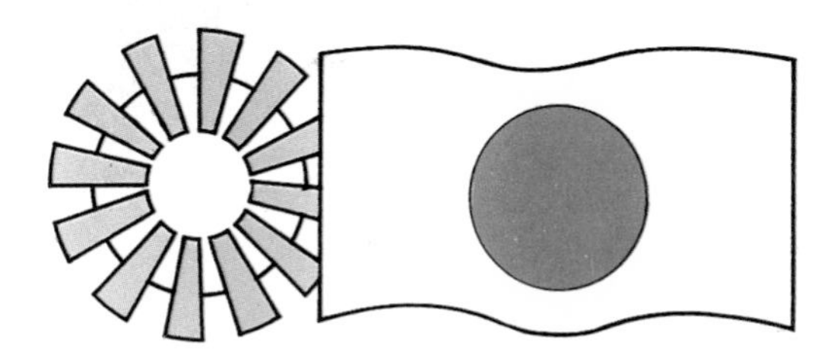

Japan is building many small wind turbines for electricity and heat in remote country places. Using small turbines is cheaper than sending electricity to these areas through the grid.

China has installed many small turbines while her neighbor, *Mongolia*, uses portable turbines. They provide nomadic shepherds with power for lights, electric fences and even hot drinks!

Glossary

Axis The shaft at the center of a wind turbine's blades. If it points upward, it is vertical; if it points along the ground, it is horizontal.

Generator The part of a power station, or wind turbine, that converts mechanical power into electricity.

Hurricane A very strong wind, which blows faster than 112 kph (70 mph).

Mill A machine for grinding crops between stones to make flour. It was among the first of all machines to be powered by the wind.

Rotor The revolving part of any machine.

Turbine Any device with blades which transforms a flow of air or water into a rotating movement.

Vane An object mounted on a swivel, which always points away from the wind like a flag. The vane at the back of a windmill or wind pump keeps the blades facing the wind.

Watt The unit for measuring electrical power. A lightbulb might use 60 or 100 watts of electricity. A kilowatt (kw) equals 1,000 watts, a megawatt (mw) one million.

Index

Acknowledgements
The publishers wish to thank the following people and organizations who have helped in the preparation of this book:
Friends of the Earth; Coventry (Lanchester) Polytechnic; Geoffrey Barnard (Earthscan); Dr S. Salter, University of Edinburgh; Energy Technology Support Unit (ETSU, Harwell); Gifford Technology; Japanese Embassy; National Engineering Laboratories, East Kilbride UK; Dr P. J. Musgrove, Reading University; Queens University, Belfast.

Typeset by Dorchester Typesetting